MODELING ATTITUDE VARIANCE FOR ACOUSTIC SIGNATURE SIMPLIFICATION IN SMALL UASS USING A DESIGNED EXPERIMENT IN A HARDWARE-IN-THE-LOOP SIMULATION

I. Introduction

Background

The advent of Radio Detection and Ranging (RADAR) technology forced acoustic aircraft detection from its position as the state of the art in aircraft detection during World War II. Since then, aircraft acoustic research has focused primarily on aircraft noise abatement studies (with a brief detour during the Vietnam War) as the planning and budgeting process was focused on peer or near-peer scenarios with RADAR being the primary aircraft detection concern. However, the dominance of the United States military's conventional capabilities has led to an evolution from conventional style warfare to the preponderance of the US's military operations being against unconventional, non-state actors in semi-hospitable environments. This has led to rapid growth of the Department of Defense's (DoD) Unmanned Aerial System (UAS) fleet as it is able to provide persistent Intelligence, Surveillance and Reconnaissance (ISR) and surgical strike capabilities without putting pilots at risk and with minimal disturbance to local populations.

The nature of these operations and the fact that non-state actors typically lack RADAR capabilities has led to a growth in interest DoD-wide for measuring and modeling acoustic aircraft signatures and providing mission planning tools related to these acoustic signatures. The focus of acoustic aircraft signature research thus far has

1

been on providing acoustic mission planning tools for preplanned routes based on possible listener positions and avoiding detection at those positions. Current research has also been effective at providing worst-case detection ranges (based on a given set of flight and weather parameters) for UAS pilots remotely flying missions that are not preplanned. However, there is a current lack of capability as it relates to providing real-time acoustic signature information to UAS pilots remotely flying missions that are not preplanned. There are a number of reasons why this capability is not currently available, but the primary reason is that running these acoustic models is computationally intensive and use of the complete, high-resolution models that are available cannot be executed in real-time. This research explores one possible way forward with achieving this real-time capability.

Brief Description of Aircraft Acoustic Models

The three primary components of an aircraft acoustic model are the source definition, the propagation model and the detection model. The source definition is the acoustic signature of the aircraft as measured by typically utilizing an array of ground microphones as the aircraft is flown overhead. The propagation model is a physical model of how the aircraft acoustic signal is attenuated as it moves through the air and towards the perspective listener. Detection models take these propagated sound levels and attempt to mimic the human auditory system in order to provide a probability of detection based on a combination of both the propagated aircraft acoustic levels and the ambient acoustic levels at the listener's position.

Since development of detection models typically lies in the realm of hearing science and the implementation of these models is not computationally intensive, the detection models were not considered for further study with this research. Likewise, acoustic propagation modeling has been studied intensively by the physics and engineering communities and most aspects of propagation are not incredibly computationally intensive. Computational intensity in this case, stems from the fact that the entire source definition (a full 360 degree sphere) is typically propagated when mission planning, which leads to lack of real-time capability. The focus of this research is to provide a methodology for simplifying the acoustic source definitions of UASs. This methodology is focused on the portion of the UAS community in which assets are tasked dynamically (without the ability to preplan missions to achieve acoustic stealth) and which should benefit from a real-time aircraft acoustic model.

Operational Assumptions and Scope

Since a preplanning (non real-time) capability already exists, the research will be geared towards dynamic mission tasking scenarios. Additionally, for situations in which acoustic stealth is desired, the UAS operator is typically made aware of one or multiple locations in which there may be listeners. Thus, this research will assume that the listener's location (point or area) is known. It is also possible that the "listener" may be something other than a human (ranging from something as unsophisticated as a trained animal to electronic listening devices). In any case, the listener is modeled by some specific detection model.

While it is possible that an aircraft (especially a small UAS designed to be quiet) may be visually detected by a spotter before it is heard, it is more often the case that visual detection of the aircraft is cued from auditory detection. Most DoD UASs designed for ISR are also typically painted in color schemes that blend with the sky when viewed from the ground and/or utilized at night with no external lights as to avoid visual detection. This research assumes that visual detection is not of concern.

Approach

Early acoustic research efforts were geared towards attempting to implement and integrate a real-time acoustic model either onto a small UAS ground station or onboard the aircraft. These efforts were focused on very tight integration with the UAS (from a time perspective) in order to acoustically propagate very few paths to the listener (based on current and future aircraft location relative to the listener). While the aircraft's position at some time in the future is easy to approximate based on heading and velocity, the attitude of the aircraft may vary significantly even within the scale of one second. As a step towards achieving a real-time model, this research focuses on modeling the UAS's maximum and minimum attitude (roll, pitch, and yaw) values as a function of select aircraft flight parameters and some environmental (weather) factors. Concepts from experimental design are utilized in order to generate these models and experimentation is conducted in a virtual simulation environment. Live, complementary flight testing was not available to complement the simulation results.

Research Objectives

The goals of this thesis include:

1. Use simulation and experimental design to develop empirical models for the minimum and maximum for each of the three aircraft attitude parameters (roll, pitch and yaw) for level flight with the chosen aircraft platform (Sig Rascal 110).

2. Extend the empirical models into models utilizing tolerance intervals for the minimum and maximum roll, pitch and yaw over the range of the independent variables.

3. Demonstrate how the models developed apply to implementation of a real-time acoustic model.

4. Propose a methodology for developing models for new aircraft platforms and for validating results with real-world flight test.

Thesis Overview

This chapter provided a brief background motivating the research, a brief description of existing acoustic models, the operational assumptions and scope of this research, as well as the approach and the objectives of the research. The next chapter reviews the literature relevant to the acoustic models this research supplements and experimental design. Chapter 3 describes the equipment, procedure and experimental design methodology used in the research. Chapter four describes the experimental results and the resulting empirical models. The final chapter provides research conclusions and recommendations for follow-on activities and research.

II. Literature Review

Chapter Overview

This chapter reviews the current state of the art relevant to acoustic modeling and motivates the need for additional research in the area. While the primary impetus guiding this effort was interest from an Air Force sponsor, it should be noted that there are other sources providing motivation for research in the chosen area.

The United States Air Force Unmanned Aircraft Systems Flight Plan 2009-2047 stresses the importance of autonomy and modularity as primary guiding principles in developing UASs in the future and highlights covert operation as one of the primary benefits of UASs [1]. This research proposes a methodology for reducing the acoustic footprint propagated to any listener while providing the UAS operator meaningful information regarding whether the UAS can be heard at specific points of interest. The methodology lends itself to the concept of a single operator controlling multiple UASs.

In a 2012 article in Armed Forces Journal, Spinetta and Cummings warn of an implicit Air Force policy change (since the departure of Defense Secretary Gates in 2011) focusing acquisition efforts back on manned platforms after a shift to unmanned aircraft during Secretary Gate's tenure [2]. The methodology explored in this research could be applied to both manned and unmanned aircraft (although manned aircraft tend to be much louder as system power requirements are greatly increased by aircrew life support systems).

The remainder of the literature search is broken out by topics relevant to the research. The primary topics covered are research relevant to aircraft acoustic source modeling and the fundamental concepts of experimental design required for this research.

Aircraft Acoustics Source Modeling

While acoustic propagation and human detection modeling is critical to the implementation of this research, the focus is on developing a methodology for reducing the area of the acoustic source propagated without reduction in the fidelity of the source model being utilized. Therefore the discussion here is limited to the background of aircraft noise models leading up to the noise source methodology this research was intended to augment.

Most research in source modeling methodologies is geared towards rotary-wing aircraft (helicopters) as their source characterizations tend to be quite complex and directional along certain azimuths. These methodologies perform well for representing acoustics sources for fixed-wing aircraft. Early efforts were geared towards noise around heliports [3] and showed that gross emissions are well modeled [4]. Efforts to more accurately represent the aircraft's acoustic signature led to a generalized source model by Moulton in 1990, in which the source was simplified to the highest sound pressure level measured from the aircraft [5]. Separate researchers also explored a representation of the polar directivity and magnitude using numeric curve fits [6], [7]. More recent research includes the addition of elevation with the polar direction in representing the noise source [8], [9]. Two models represent the current state of the art in providing three-dimensional source representations. The first is a model developed by National Aeronautics and

Space Administration (NASA) in conjunction with Wyle Laboratories involving storing grid-spaced acoustic measurements and utilizing interpolation algorithms for filling in the gaps [8]. The second is a model developed by the Swiss Federal Laboratories for Materials Testing and Research utilizing a spherical harmonic (SH) representation which relies on a least-squares analysis to determine the coefficients of the SH expansion [9]. This research was conducted with the SH approach in mind, but could also be beneficial if utilized in conjunction with the NASA interpolation model.

Experimental Design

Montgomery defines an experiment as "a test or series of runs in which purposeful changes are made to the input variables of a process or system so that we may observe and identify the reasons for changes that may be observed in the output response" [10]. For this research, we want to determine what factors (input variables), if any, affect the attitude variability of an aircraft (output response).

The statistical field of Design of Experiments describes the process of constructing efficient and effective experiments. In contrast, naïve experimentation may lead to inefficiencies such as varying one factor at a time (OFAT) or choosing inputs that are linearly related. The primary issue with OFAT experimentation is that it does not consider situations in which two or more factors have an interaction effect on the response. Additionally, choosing correlated levels for multiple factors results in multicollinearity which can cause problems such as model misspecification or large variances and covariances for the regression coefficients. A good way to avoid multicollinearity is to use orthogonal, factorial designs. Orthogonality is achieved by setting the input factors at coded levels of -1 and 1 representing the minimum and maximum factor values you are interested in observing. A factorial design is a design in which each possible factor combination is explored for a total of 2^k experimental runs with k being the number of input variables of interest. All of the factors being studied in this research are quantitative, simplifying some of the discussion as it relates to experimental design [10].

There are three basic principles in experimental design: randomization, replication and blocking. Randomization is important since it reduces the effect of factors that have not been explicitly included in the experimental design. Randomization also validates the assumption (required by the underlying statistical methodology) that the experimental observations be independently distributed random variables. All the experiments conducted in this research were randomized. Replication is the experimental repetition of factor combinations and is important since it provides the experimenter with a true estimate of the experimental error, which is used as comparison for determining the statistical significance of the terms in a statistical model. In the case of this research, replicated center point runs (coded value of 0) were used. Additionally, with a 2^k factorial design, if one or more factors are determined to be insignificant, the design "collapses", forming a replicated factorial design in the lower number of factors. Blocking is a technique for eliminating variability from nuisance factors but was not used in this research as the experiments were simulations and nuisance factors were not identified [10].

Once the data are collected, a model for the data is constructed. The standard methodology for building models relating input variables to their output response involves multiple linear regression and the method of least squares for estimating the regression coefficients.

Another important concept in experimental design, is model adequacy checking. Using the multiple linear regression model carries several assumptions. These assumptions are that: 1) the relationship between the response and the input variables is at least approximately linear, 2) the residual errors (that is the difference between each of the observations and the fitted model) have a mean of zero, 3) the residual errors have constant variance, 4) the residual errors are uncorrelated, and 5) the residual errors are normally distributed. These assumptions should be examined anytime least squares is used to make statistical inferences and are typically checked using various plots of the residual errors. Plotting the residuals versus the fitted values provides a good test for assumption 3. A plot of the residuals in time sequence is useful in determining whether assumption 4 holds. Assumption 5 is checked by plotting the residuals in a normal probability plot and ensuring they are at least approximately normally distributed. A reasonable test is called the fat pencil test: if a fat pencil can be laid along the normal probability plot of the residuals and cover the residuals, the normality assumption is assumed to be met [11].

Finally, the experimental data is also examined to determine if there are any outliers. The primary diagnostic for identifying data outliers is to scale the residuals so that they should typically fall within a certain range independent of the experimental data utilized so that the same criteria can be applied from model to model. With the residuals used in this analysis, any data points with scaled residual values whose absolute values are near or above three should be closely scrutinized to determine if there are issues.

Outliers may indicate problems with the experimental data and can either be taken out of the model, remain in the model, or new data can be collected to replace outlying data point [10].

Summary

The literature review motivates research into providing improved real-time acoustic information for small UAVs. The history of acoustic signature directionality in rotary-wing aircraft characterization is examined, which led to the development of two high-fidelity methodologies specifically supplemented by this research. Finally, an overview of experimental design was provided along with a brief explanation of some of the underlying experimental design concepts critical in this effort.

III. Methodology

This chapter describes the process utilized to meet the objectives of this research. To meet these objectives, data are collected using simulation and the data are analyzed using various statistical methods. The first section describes the hardware, software and processes utilized for the hardware-in-the-loop simulation. The second section discusses the experimental design utilized. The third section discusses the resulting of the tolerance intervals.

Simulation Hardware, Software and Processes

Typical components of a SUAS include the air vehicle, payload, ground station, communications, launch and recovery hardware and ground support equipment [12]. Since this research is limited to simulation modeling, the physical air vehicle (including payload and launch and recovery hardware) is not required. However, the hardware-in-the-loop simulation utilizes the ground station (with a few modifications from the real-world flight configuration) and the autopilot. These components and their configurations are discussed below including a brief overview of the air vehicle for completeness.

Air Vehicle

The air vehicle simulated in this experiment is the Sig Rascal 110, a small (110" wingspan), widely-available, hobbyist RC aircraft. This air vehicle was chosen because there is a simulation model available to use with the simulated flight environment chosen (FlightGear). Additionally, should a follow-on validation be possible, the AFIT SUAS program has both gas and electric variants of the Sig Rascal 110 available for flight

12

testing and each are approved for USAF test on the range at Camp Atterbury, Indiana.

Figure 1 shows the Sig Rascal 110 on the runway at Camp Atterbury.

Figure 1. Sig Rascal 110

Autopilot

While there are many commercial-off-the-shelf (COTS) autopilot alternatives varying widely in cost and capability, the ArduPilot Mega (APM) version 2.6 is the autopilot used in this research, and for the majority of research conducted at AFIT. The ArduPilot is a low-cost autopilot based on the Arduino open-source electronics prototyping platform and utilizes an Inertial Measurement Unit with an array of accelerometers, gyroscopes and magnetometers for navigation. The APM works with ground vehicles as well as fixed and rotary wing aircraft depending on the firmware that is loaded on the APM. The APM also has the capability to attach peripherals such as a modem (for control and telemetry), a global positioning system receiver, and a barometric pitot-static tube. The ArduPilot is itself an open source project thus lending itself to easy code modification which is often necessary in research [13].

The APM was chosen primarily for its ability to run with the chosen software (FlightGear and Mission Planner) as a hardware-in-the-loop simulation. Additionally, the APM is similar in processing power and flight functionality to autopilots used in many currently fielded systems [12], which is important if the results of this research are to be

applied to other SUAS platforms. For this research, the APM was connected directly to

the computer via USB (with no other peripherals attached), and any data received from

the internal sensors as well as peripherals attached (GPS module and pitot-static tube)

was simulated using the flight simulation environment (Flight Gear). Figure 2 is a

picture of the APM 2.6.

Figure 2. ArduPilot Mega Version 2.6

Ground Control Station

The Ground Control Station utilized for this research is a standard PC laptop

running Microsoft Window 7. During real-world operations, the laptop would typically

run only the Mission Planner software and be configured with a single wireless modem

connected via USB for two-way communication with the aircraft autopilot. Mission

Planner is an open source software platform used to monitor the operating vehicle's

status as well as plan, save and load autonomous missions into the autopilot either before

or during flight. In addition, Mission Planner is used to load firmware to the autopilot,

setup, configure and tune the autopilot, record detailed telemetry logs, and view and analyze the telemetry logs. Most important for translating results to military application, Mission Planner's functionality is comparable to that of most similar fielded SUASs. The configuration for running a hardware-in-the-loop simulation requires a flight simulation environment (in this case, FlightGear) and connecting the autopilot to the PC via the USB connection. A wireless modem is not required for HIL simulation. A screen capture of the typical Mission Planner environment is shown in Figure 3. Detailed specifications for the hardware and software versions utilized are in Appendix A.

Figure 3. Mission Planner Screenshot

Flight Simulation Environment

To develop appropriate attitude variance models, data points with a variety of wind speeds and wind headings (relative to aircraft direction) are collected. This task is very difficult (and relatively expensive) to achieve with real-world flight test. Therefore this research uses the HIL simulation with FlightGear software providing the simulated environment. FlightGear is an open source flight simulator "created to provide a

15

sophisticated and open framework for use in research/academia, pilot training, as an industry engineering tool, for do-it-yourselfers to pursue their favorite interesting flight simulation idea, and…as a fun, realistic, and challenging desktop flight simulator [14]." FlightGear utilizes one of three flight dynamics models determined by the format of aircraft model being utilized. In this case, the Sig Rascal 110 flight dynamics model was created using JSBSim which is an open source, six degrees of freedom library (written in C++) for simulating flight dynamics and control of the aircraft. Aircraft are modeled by collecting and storing mass, aerodynamic and flight control properties in an XML configuration file [14]. The communications architecture utilized for these HIL simulations is illustrated in Figure 4. The APM navigation logic used in HIL simulation is the same as real-world since the navigation processes use simulated aircraft sensor and positional data in exactly the same manner it is used in real-world flight. This ensures that (assuming the simulation environment and the aircraft model are accurate) simulation results are transferrable to real world flight. Additionally, an optional remote control (R/C) transmitter and receiver were not used but are helpful for troubleshooting, transitioning between test points and for tuning the hardware for real-world flight test. Details on specific flight simulation software versions utilized for this research are given in Appendix A.

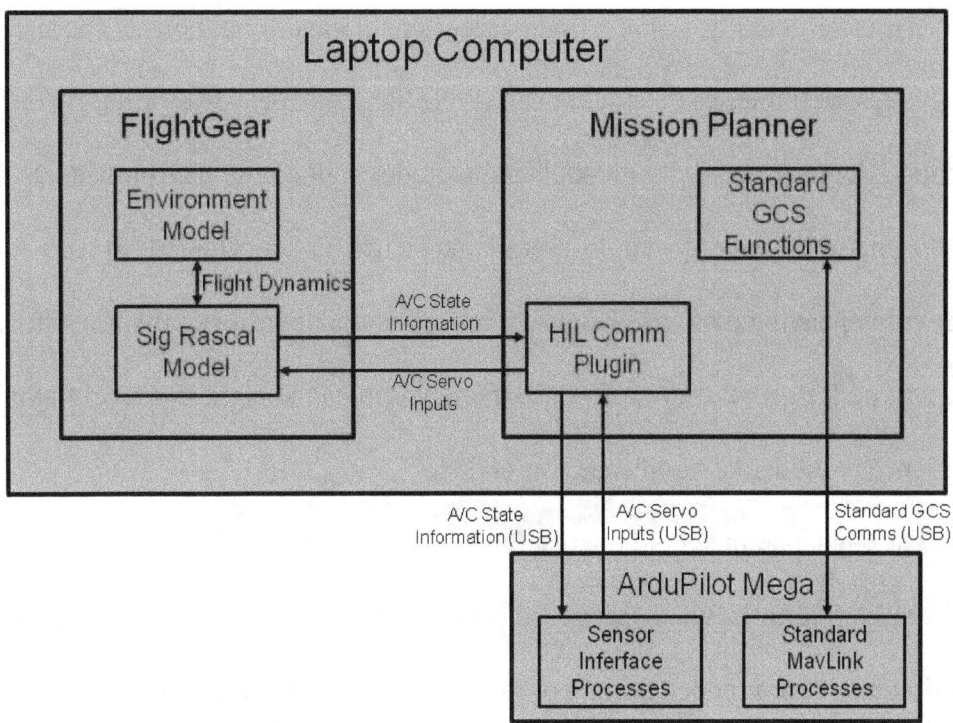

Figure 4. Hardware-in-the-Loop Architecture

HIL Procedures

The first step in conducting a HIL simulation is to load the HIL version of the fixed-wing firmware to the autopilot using Mission Planner. Of note, many other AFIT researchers have experienced compatibility and stability issues with certain combinations of HIL firmware, Mission Planner and FlightGear software packages (as can be expected in using several different open source software packages). While the utilized configuration was sometimes difficult to initiate, stability and compatibility were not an issue once the simulation was running. See Appendix A for the software and firmware versions utilized in this research.

When the HIL firmware is installed, generic fixed-wing tuning settings are loaded to the APU. By changing these tuning parameters, the autopilot is configured to work

17

effectively and efficiently with the flight characteristics and limitations of the airframe. Since this research focuses on straight and level flight, the focus was primarily on correcting issues that existed with pitch and altitude oscillations. The procedures from the Total Energy Control System for Speed and Height Tuning Guide [15] were followed and the corresponding detailed tuning settings are available in Appendix A. After the tuning was performed, the aircraft was extremely stable in straight and level flight until the presence of moderate turbulence was added.

The flight parameters of interest (and thus changed) in this research are aircraft altitude, aircraft throttle, wind speed, and wind heading relative to aircraft heading. All of the simulation runs modeled flight over the Pacific Ocean so the altitude in above ground level (AGL, which is what is utilized by Mission Planner based on the home location) is approximately equivalent to altitude mean sea level (MSL). The MSL is measured in meters and is easily changed in Mission Planner. The aircraft throttle settings use throttle percentage and also easily changed in Mission Planner. Conducting the runs was accomplished by setting a waypoint heading west over the Pacific (having the aircraft flying on one straight flight path) and varying the wind speed and direction relative to that flight path for each of the test points. Wind speed is measured in knots and is changed in the weather menu in FlightGear along with the wind direction (which is measured in degrees). Additionally, turbulence is also adjusted from this window. Figure 5 shows the FlightGear weather dialog in which these parameters are changed.

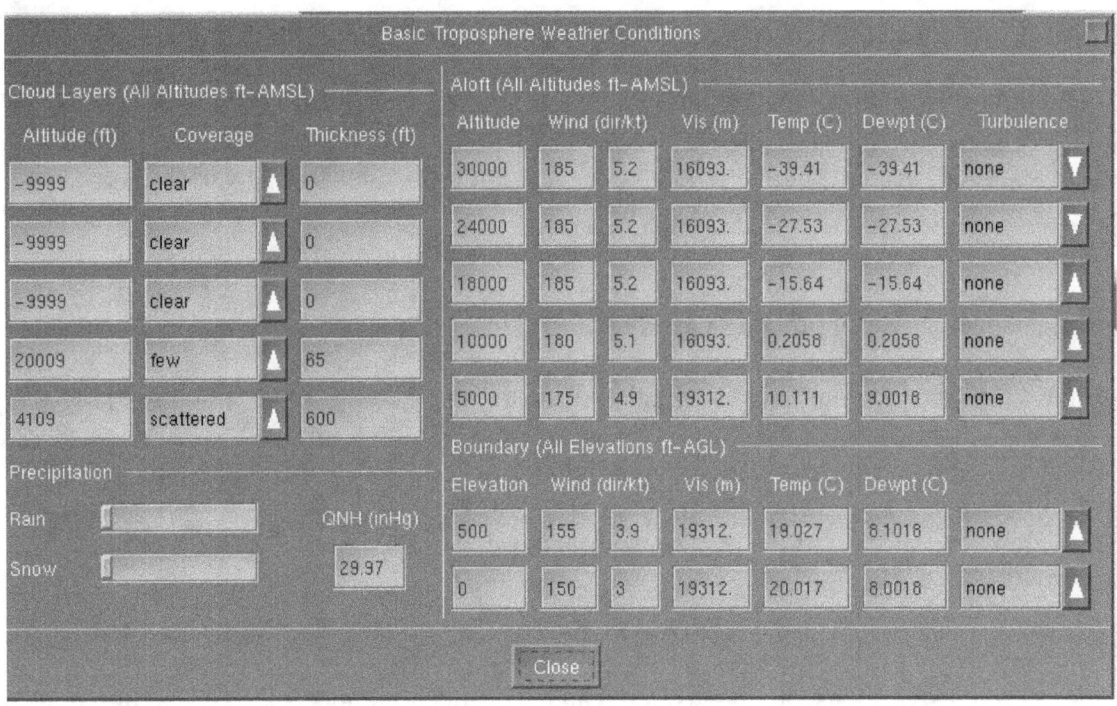

Figure 5. FlightGear Weather Dialog Box

Experimental Design and Data Collection

Due to the nature of the variables of interest, a second order model was deemed

likely needed to accurately model the attitude variance of the aircraft. The relative

heading variable was the primary driver behind this decision. Often with experimental

design, a screening design is used to determine which factors affect the response along

with some center point runs to test for lack of fit (and determine if a higher order model is

required). In this research, the relative heading to varied between 0 and 180 degrees and

we assumed the response would be symmetric for values between 180 and 360 degrees.

Additionally, it was reasoned that (at least some of) the attitude responses would be

nonlinear in moving from 0 to 180 degrees and that adding the runs required to estimate

second order effects would only require two additional runs per factor (for a total of eight runs). A central composite design consisting of a 24 factorial design augmented with eight axial runs and four center point runs for a total of 28 runs was used. Table 1 shows the non-randomized design with both coded and natural values for the variables. Of note, to ensure the variance of the predicted response depends only on the distance from the design center (a useful property called rotatability), by Equation 1, with the coded value of the axial runs as α and F as the number of factorial runs (2^4), α = 2. Since the range of the factors is based on the operational limitations of the aircraft, the choice of α determines the spacing of the experimental factor levels.

$$\alpha = \sqrt[4]{F} \qquad\qquad (1)$$

Wind speed was varied between 0 and 16 knots; 15 knots is the typical maximum wind speed for AFIT SUAS operations and the 0 to 16 range allows for equally spacing five wind speed levels to be 0, 4, 8, 12, and 16 knots. Relative heading was set at values of 0, 45, 90, 135 and 180 degrees. The maximum throttle setting is 100% and the typical minimum for this aircraft to stay aloft is about 40% so throttle settings were set as 40, 55, 70, 85 and 100%. Since the AFIT SUAS program typically operates between 100 and 1,000 feet AGL, levels used were 30, 105, 180, 255 and 330 meters.

Table 1. Non-randomized Design with Coded and Natural Values

	Coded Values					Natural Values			
Run	Wind Speed	Relative Heading	Throttle	Altitude		Wind Speed	Relative Heading	Throttle	Altitude
1	1	1	1	1		12	135	85	255
2	1	1	1	-1		12	135	85	105
3	1	1	-1	1		12	135	55	255
4	1	1	-1	-1		12	135	55	105
5	1	-1	1	1		12	45	85	255
6	1	-1	1	-1		12	45	85	105
7	1	-1	-1	1	2^4 Factorial	12	45	55	255
8	1	-1	-1	-1		12	45	55	105
9	-1	1	1	1		4	135	85	255
10	-1	1	1	-1		4	135	85	105
11	-1	1	-1	1		4	135	55	255
12	-1	1	-1	-1		4	135	55	105
13	-1	-1	1	1		4	45	85	255
14	-1	-1	1	-1		4	45	85	105
15	-1	-1	-1	1		4	45	55	255
16	-1	-1	-1	-1		4	45	55	105
17	2	0	0	0		16	90	70	180
18	-2	0	0	0		0	90	70	180
19	0	2	0	0		8	180	70	180
20	0	-2	0	0	Axial Runs	8	0	70	180
21	0	0	2	0		8	90	100	180
22	0	0	-2	0		8	90	40	180
23	0	0	0	2		8	90	70	330
24	0	0	0	-2		8	90	70	30
25	0	0	0	0		8	90	70	180
26	0	0	0	0	Center points	8	90	70	180
27	0	0	0	0		8	90	70	180
28	0	0	0	0		8	90	70	180

A couple of other considerations went into executing the design. First of all, one of the three basic principles of experimental design is randomization (along with replication and blocking), so each iteration of this experiment was randomized [10]. Additionally, as a best practice, when factor values remained the same between runs, the factor values were reset and verified. The responses were chosen as the maximum and minimum values for each of yaw, pitch and roll (total of six responses). Because the amount of time spent at each point may have some effect on the outcome, each design point was held for two minutes once the aircraft got to the proper altitude and held steady at that altitude.

As soon as an autopilot connects to Mission Planner, Mission Planner begins to log (all types of) data in .tlog files. These files were converted to a usable format using the tlog Extractor utility [16] so that time, roll, pitch and yaw could be extracted from the two minute blocks of the telemetry data. Of note, since the average yaw changes based on the aircraft heading (pitch and roll always stay around zero degrees for straight and level flight regardless of heading) 270 degrees was subtracted from the maximum and minimum headings to account for the average yaw of 270 degrees since the aircraft heading was due west.

Tolerance Interval

Tolerance intervals are used in this research as a prediction of the maximum and minimum roll, pitch and yaw. They are used to provide assurance that the aircraft's attitude will not vary outside of the determined maximum and minimum bounds while in flight. The tolerance interval is a statistical interval bounding an arbitrary sequence of future samples. Tolerance intervals require a desired population proportion for the interval to bound (indicated by p in Equations 3 – 5) as well as confidence level (indicated by γ in Equations 3 – 5) to define the interval [17]. To compute the tolerance interval, first the point estimate is computed according to Equation 2. Next, a value for k_1 is computed according to Equations 3 – 5, with z_p and z_γ being the z-scores determined from the chosen values of p and γ, and the N being the number of samples

used to generate the intervals. One-sided tolerance intervals are determines using
Equations 6 and 7 with s being the standard error computed at the design point of interest
[17]. Lower tolerance bounds will be used for minimum value responses (Equation 6)
and upper tolerance bounds will be used for maximum value responses (Equation 7).
Those computed tolerance interval points will be fit using linear regression and the
significant factors from the models fit previously and used to compute the tolerance
intervals. This would result in separate models for each percent tolerance interval that
may be utilized, but will simplify utilization of the tolerance intervals as only a point
estimate will need to be computed.

$$\hat{y}(x_0) = x_0' b \tag{2}$$

$$a = 1 - \frac{z_\gamma^2}{2(N-1)} \tag{3}$$

$$b = z_p^2 - \frac{z_\gamma^2}{N} \tag{4}$$

$$k_1 = \frac{z_p + \sqrt{z_p^2 - ab}}{a} \tag{5}$$

$$Y_L = \hat{y} - k_1 s \tag{6}$$

$$Y_U = \hat{y} + k_1 s \tag{7}$$

Real World Validation Plan

Real-world validation of the models, while planned, was not conducted due to
problems with the aircraft autopilot unit and an inability to reschedule the validation runs.

An extensive amount of the research time allocated for this effort did not produce results because of these issues.

The original plan was to utilize a Super Sky Surfer (an inexpensive, foam, hobbyist UAS) for the real world validation and utilize a flight dynamic model for a very similar foam aircraft within FlightGear to develop the attitude variability models. A Super Sky Surfer was built solely for this effort and acoustic measurements were made in AFRL's anechoic chamber. In addition to developing and validating the attitude variability models, the range was equipped with precision digital sound level meters and the flight test plan (consisting of 34 data points) would have provided additional acoustic data for other aspects of this research that were not realized. Ultimately, many hours of research were abandoned in order to complete one aspect of the planned research. Also at that point, the decision was made to utilize the Sig Rascal 110 since AFIT owns several different variations of the aircraft and real-world validation of the same model used for simulation (in the future) should provide more compelling results.

IV. Analysis and Results

The analysis and results chapter describes the analysis process for the data collecting utilizing the methodology from Chapter 3 and provides an interpretation of that data. First, the data from the experimental design without turbulence is briefly examined, followed by an examination of the data collected when adding turbulence to the simulation environment (FlightGear). Finally, the results are examined utilizing the tolerance interval methodology.

General Notes on Regression Analysis

JMP version 10 (developed by the SAS Institute) was used to fit the experimental data to their respective regression models. Two model building methodologies were utilized. One method fit a full quadratic model of the regressors sorting the parameter estimates by p-values to help determine which regressors to include in the final model. With this methodology, enough degrees of freedom exist to compute the parameter estimates for each of the quadratic terms simultaneously but higher order interactions cannot be examined. The second methodology utilizes the JMP screening tool, which allows one to examine higher order interactions using a variety of tools including the half-normal probability plot of the regression term's contrast. This methodology allows you to look at the higher order terms that may actually represent the real-world system although one has to be careful with the alias structure and avoid over fitting the data. With these two methodologies, regressors were added or removed by considering a combination of improvement in the adjusted R^2 value, achieving $\alpha \approx .05$ level of significance for the regressors, achieving a parsimonious model with little correlation between the regressors and maintaining model hierarchy where appropriate. In the analyses that follow, the chosen models relied on one of two methods, so either the sorted parameter estimates are utilized for method 1, or the half-normal probability plot are utilized for method 2.

Modeling Straight and Level Flight Without Turbulence

The initial experiment was conducted without turbulence in the simulation environment. This experimental design with the collected responses is provided in Table 2. Maximum and minimum yaw values were computed by subtracting 270 degrees (actual heading) from the measured maximum and minimum yaw values. All of the measured responses (minimum and maximum for roll, pitch and yaw, orientation shown in Figure 6) are given in degrees. Of note, the maximum yaw for run number 18 is much larger than other maximum yaw values also making the range (difference between the maximum and minimum yaw) much larger than for the other runs. Additionally, the ranges between the maximum and minimum values for each of roll, pitch and yaw (roll and pitch especially) appear to be smaller than expected; this potential issue is addressed later.

A summary of results for the runs with no turbulence are given in Table 3. Detailed results for this modeling effort are provided in Appendix B, but are not pertinent to the discussion here since these models are not being recommended for utilization.

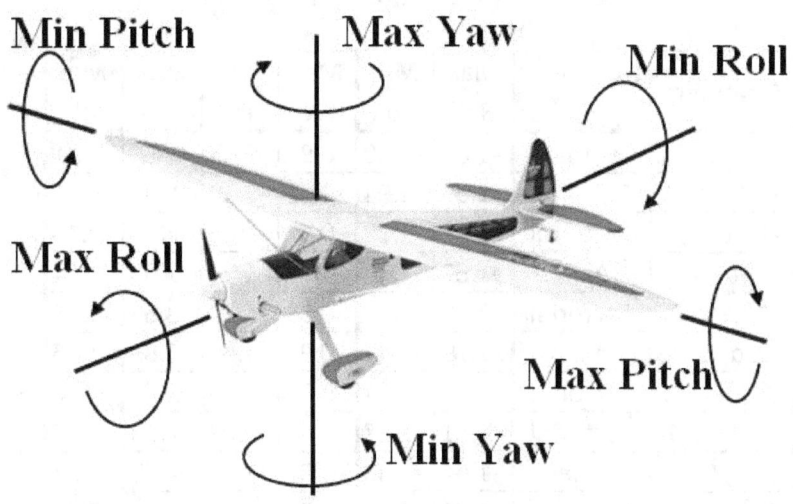

Figure 6. Depiction of Max and Min Roll, Pitch and Yaw

Table 2. Observed Responses for Flight Without Turbulence

Run Number	Treatment	Treatment Type	Max Roll	Min Roll	Max Pitch	Min Pitch	Max Yaw	Min Yaw
1	0, 0, 0, 0	Center Pt	-1.81	-2.50	-0.90	-1.43	8.42	7.70
2	1, -1, -1, -1	Factorial	-1.63	-1.94	-0.51	-1.10	10.45	10.27
3	-1, -1, -1, 1	Factorial	-1.72	-1.89	-0.52	-1.06	-2.97	-3.13
4	0, -α, 0, 0	Axial	-1.67	-2.44	-0.96	-1.29	-1.06	-1.64
5	1, 1, 1, 1	Factorial	-2.18	-2.97	-0.85	-1.62	-3.68	-4.91
6	0, α, 0, 0	Axial	-1.79	-2.43	-0.93	-1.64	-0.83	-1.56
7	-1, 1, -1, -1	Factorial	-1.42	-2.74	-0.78	-1.13	3.87	3.04
8	-1, 1, -1, 1	Factorial	-1.81	-1.97	-0.87	-1.02	-2.10	-2.41
9	0, 0, 0, α	Axial	-1.64	-2.43	-0.87	-1.64	-4.59	-5.28
10	1, -1, -1, 1	Factorial	-1.71	-2.27	-0.55	-1.45	-4.21	-5.21
11	0, 0, α, 0	Axial	-2.30	-2.76	-1.05	-1.91	7.64	7.11
12	-1, -1, 1, 1	Factorial	-2.24	-2.81	-0.91	-1.57	-2.26	-2.64
13	0, 0, 0, 0	Center Pt	-1.83	-2.42	-0.96	-1.81	8.40	7.65
14	-1, -1, 1, -1	Factorial	-2.09	-2.95	-1.05	-1.42	3.79	3.04
15	-1, 1, 1, 1	Factorial	-2.33	-2.48	-0.91	-1.61	-2.32	-2.54
16	1, -1, 1, -1	Factorial	-1.79	-2.97	-0.93	-1.52	9.47	8.52
17	0, 0, 0, 0	Center Pt	-1.97	-2.55	-0.92	-1.56	8.64	7.72
18	1, 1, -1, -1	Factorial	-1.20	-2.53	-0.37	-1.30	50.18	9.61
19	0, 0, 0, 0	Center Pt	-1.82	-2.47	-0.95	-1.41	8.41	7.74
20	-1, 1, 1, -1	Factorial	-1.79	-3.20	-0.78	-1.49	4.37	2.86
21	α, 0, 0, 0	Axial	-1.73	-2.58	-0.97	-1.70	15.54	15.09
22	0, 0, 0, -α	Axial	-1.68	-2.59	-0.83	-1.43	9.52	8.72
23	-α, 0, 0, 0	Axial	-1.71	-2.44	-0.69	-1.59	1.22	0.59
24	-1, -1, -1, -1	Factorial	-1.74	-2.02	-0.79	-1.11	4.15	3.58
25	1, -1, 1, 1	Factorial	-2.31	-2.61	-1.26	-1.49	-4.41	-4.75
26	0, 0, -α, 0	Axial	-1.32	-1.84	-0.41	-1.17	8.92	8.34
27	1, 1, 1, -1	Factorial	-1.84	-3.22	-0.72	-1.68	9.49	8.69
28	1, 1, -1, 1	Factorial	-1.80	-2.35	-0.66	-1.25	-3.88	-4.46

Table 3. Summary of Modeling Results for Flight Without Turbulence

	Max Roll	Min Roll	Max Pitch	Min Pitch	Max Yaw	Min Yaw
R^2	0.776	0.781	0.801	0.634	0.508	0.898
R^2 Adj	0.737	0.743	0.701	0.588	0.447	0.869
RMS Err	0.143	0.186	0.112	0.151	7.894	2.161
Mean	-1.816	-2.513	-0.817	-1.443	5.006	2.919
Model Fval	<.0001	<.0001	0.0001	<.0001	0.0006	<.0001
LoF Fval	0.0225	0.1607	0.0153	0.1565	0.8488	<.0001
Regression Equation Terms with p-values[*]	RH - .39	RH - .04	WS - .72	WS - .11	WS - .037	WS - .0004
	Thr - <.0001	Thr - <.0001	RH - .24	RH - .16	Alt - .0004	RH - .9730
	Alt - .0015	Alt - .011	Thr - <.0001	Thr - <.0001	WS * Alt .0365	Alt - <.0001
	RH * Alt - .06	RH * Alt - .039	Alt - .23			RH * RH - <.0001
			WS * RH - .11			WS * Alt - 0.001
			WS * Thr - .04			Alt * Alt - .0004
			RH * Thr - .015			
			Thr * Thr - .037			
			WS * Alt - .04			
* Alt - Altitude, RH - Relative Heading, Thr - Throttle, WS - Wind Speed						

All of the models for the no turbulence flights were fit using the sorted parameter estimates (method 1). All of the models except perhaps the maximum yaw and minimum pitch models provided reasonable R^2 values, so the models do a reasonable job of explaining the variation of the system. Additionally, all of the models show that they are significant at better than the $\alpha = 0.05$ level. However, the lack of fit test is a concern for three of the six models, and is borderline for two of the models (both of which at approximately 0.16). While the lack of fit test interpretation here is difficult (due to higher order terms already being included in the model) and could be due to a small pure error sum of squares more than a real lack of fit, it may indicate that there are issues with the model. While it was promising that each of the model pairs (two each for roll, pitch and yaw) contained many of the same regression terms, the results of this experiment seemed problematic. It was theorized that the problems with the models may be due to the small variation in the responses. Table 4 below illustrates this small variation in response for roll and pitch. Of note, the maximum yaw possible outlier (Run 18) was not included in this calculation. Additionally, while the maximum roll values are biased towards negative values since the wind is always coming from the right side of the aircraft, one would expect to see at least some maximum roll values as positive values. To introduce more variation into the system (likely a closer resemblance to real-world), new models were developed involving introducing turbulence into the simulation environment.

Table 4. Range of Responses Without Turbulence

	Max Roll	Min Roll	Max Pitch	Min Pitch	Max Yaw	Min Yaw
Max Response	-1.2	-1.8	-0.4	-1.0	10.5	15.1
Min Response	-2.3	-3.2	-1.3	-1.9	-4.6	-5.3
Response Range	1.1	1.4	0.9	0.9	15.0	20.4

Modeling Straight and Level Flight With Turbulence

Four turbulence levels are available within the FlightGear simulation environment: none, light, moderate, and heavy. The intention was to choose the maximum amount of turbulence that would be indicative of real-world flight, thus inducing the most variability within reason. The setting that was chosen was the moderate setting as this small airframe in not designed to be flown in heavy turbulence. This was apparent in trial simulation runs as the flight was erratic and unstable. The measured responses for the experiment with turbulence are shown in Table 5.

Table 5. Measured Responses for Flight With Turbulence

Run Number	Treatment	Treatment Type	Max Roll	Min Roll	Max Pitch	Min Pitch	Max Yaw	Min Yaw
1	0, 0, 0, 0	Center Pt	1.25	-5.09	0.81	-2.46	9.85	4.45
2	1, 1, -1, -1	Factorial	4.59	-6.58	3.61	-5.44	15.30	4.84
3	0, 0, -2, 0	Axial	1.39	-5.19	0.72	-3.37	10.45	4.57
4	0, 0, 2, 0	Axial	0.03	-5.34	0.68	-3.49	8.45	4.26
5	0, 0, 0, 2	Axial	-1.69	-2.38	-0.34	-1.60	-4.59	-5.21
6	-1, 1, -1, 1	Factorial	-0.26	-2.80	-0.19	-2.23	-2.07	-4.46
7	-1, -1, 1, -1	Factorial	-0.68	-4.08	-0.35	-2.52	4.62	2.13
8	1, 1, 1, 1	Factorial	1.99	-6.80	1.10	-3.58	-0.43	-8.13
9	1, -1, 1, 1	Factorial	0.54	-5.40	1.48	-4.35	-3.19	-7.42
10	1, -1, -1, -1	Factorial	3.66	-7.24	3.33	-4.31	16.50	5.74
11	0, 2, 0, 0	Axial	0.98	-5.68	1.37	-2.50	1.80	-4.67
12	-1, -1, -1, -1	Factorial	0.02	-3.57	0.12	-1.89	5.75	2.38
13	0, 0, 0, 0	Center Pt	-0.12	-4.48	1.25	-3.25	8.63	5.24
14	0, 0, 0, 0	Center Pt	1.18	-4.50	0.94	-2.82	9.34	4.76
15	1, 1, 1, -1	Factorial	2.00	-7.18	2.68	-5.98	12.73	4.35
16	-1, 1, 1, 1	Factorial	-0.92	-3.94	-0.24	-2.87	-1.59	-4.12
17	2, 0, 0, 0	Axial	4.48	-7.61	3.73	-5.77	19.24	9.07
18	-2, 0, 0, 0	Axial	-1.72	-2.43	-0.80	-1.46	1.57	0.95
19	-1, -1, 1, 1	Factorial	-1.41	-3.40	-0.38	-2.37	-2.23	-3.97
20	1, -1, 1, -1	Factorial	2.61	-5.74	1.38	-4.19	11.44	5.79
21	1, -1, -1, 1	Factorial	1.14	-4.67	2.14	-3.56	-3.54	-8.84
22	0, 0, 0, 0	Center Pt	0.48	-4.49	0.89	-3.16	9.50	4.57
23	1, 1, -1, 1	Factorial	2.01	-5.48	2.69	-3.79	0.14	-8.49
24	0, -2, 0, 0	Axial	1.42	-5.26	0.34	-2.41	0.24	-3.74
25	-1, 1, 1, -1	Factorial	-0.90	-3.81	-0.19	-2.42	4.44	1.94
26	-1, 1, -1, -1	Factorial	-0.04	-3.50	0.22	-2.18	5.79	2.14
27	0, 0, 0, -2	Axial	0.10	-4.37	1.73	-3.89	11.95	6.88
28	-1, -1, -1, 1	Factorial	-0.91	-2.97	-0.14	-1.66	-3.07	-4.84

Upon reviewing the results from the new experiment, it was immediately apparent that adding turbulence increased response variability (as expected). Of note, some of the

31

maximum roll values are still negative which may be due to the fact that the wind is always coming from the right side of the aircraft. Table 6 shows the maximum, minimum and range values for the experiment with turbulence as well as a comparison to the experiment without turbulence. Of note, ranges for each of the responses increased (as expected) except for minimum yaw range. Upon examination, the culprit for the unexpected minimum yaw ranges appears to be that the maximum response for minimum yaw decreased from 15.1 in the experiment without turbulence to 9.1 in the experiment with turbulence.

Table 6. Turbulence Responses Compared to W/O Turbulence Range

	Max Roll	Min Roll	Max Pitch	Min Pitch	Max Yaw	Min Yaw
Max Response	4.6	-2.4	3.7	-1.5	19.2	9.1
Min Response	-1.7	-7.6	-0.8	-6.0	-4.6	-8.8
Response Range	6.3	5.2	4.5	4.5	23.8	17.9
Range - No Turb	1.1	1.4	0.9	0.9	15.0	20.4

This revelation led to further examination of the maximum and minimum responses between the with and without turbulence experiments. Table 7 shows the minimum and maximum yaw responses along with a new response, range, as well as the minimum, maximum and range of values for each of responses. The table shows that while the minimum and maximum values in moving from the with and without turbulence experiments did not change significantly, the range between the minimum and maximum values for each run appears to have significantly increased in adding turbulence. This may be due to the fact that roll and pitch are quickly "corrected" by the autopilot when they deviate from equilibrium, while the yaw is a navigational computation based in part on the wind speed and direction that the aircraft computes based on various sensor information collected. This difference may indicate the need to

evaluate range as a third response (and separate methodology) for yaw.

Table 7. Min, max and range responses for Yaw

Run Number	Without Turbulence			With Turbulence		
	Max Yaw	Min Yaw	Yaw Range	Max Yaw	Min Yaw	Yaw Range
1	8.42	7.70	0.73	9.85	4.45	5.39
2	10.45	10.27	0.19	15.30	4.84	10.46
3	-2.97	-3.13	0.16	10.45	4.57	5.87
4	-1.06	-1.64	0.58	8.45	4.26	4.19
5	-3.68	-4.91	1.23	-4.59	-5.21	0.62
6	-0.83	-1.56	0.73	-2.07	-4.46	2.39
7	3.87	3.04	0.82	4.62	2.13	2.49
8	-2.10	-2.41	0.31	-0.43	-8.13	7.70
9	-4.59	-5.28	0.69	-3.19	-7.42	4.23
10	-4.21	-5.21	1.00	16.50	5.74	10.76
11	7.64	7.11	0.53	1.80	-4.67	6.47
12	-2.26	-2.64	0.38	5.75	2.38	3.37
13	8.40	7.65	0.75	8.63	5.24	3.38
14	3.79	3.04	0.75	9.34	4.76	4.58
15	-2.32	-2.54	0.22	12.73	4.35	8.38
16	9.47	8.52	0.94	-1.59	-4.12	2.53
17	8.64	7.72	0.92	19.24	9.07	10.17
18	50.18	9.61	-	1.57	0.95	0.62
19	8.41	7.74	0.67	-2.23	-3.97	1.74
20	4.37	2.86	1.51	11.44	5.79	5.64
21	15.54	15.09	0.45	-3.54	-8.84	5.30
22	9.52	8.72	0.80	9.50	4.57	4.93
23	1.22	0.59	0.63	0.14	-8.49	8.64
24	4.15	3.58	0.57	0.24	-3.74	3.99
25	-4.41	-4.75	0.34	4.44	1.94	2.50
26	8.92	8.34	0.58	5.79	2.14	3.65
27	9.49	8.69	0.80	11.95	6.88	5.06
28	-3.88	-4.46	0.58	-3.07	-4.84	1.76
Max Response	10.45	15.09	1.51	19.24	9.07	10.76
Min Response	-4.59	-5.28	0.16	-4.59	-8.84	0.62
Response Range	15.04	20.36	1.35	23.83	17.92	10.14

Modeling Responses with Turbulence

Each of the seven responses measured for flight with turbulence were modeled using JMP 10 and a summary of the results are given in Table 8. All of the models except the minimum and maximum yaw models were developed using the sorted parameter estimates of the full quadratic model fits (method 1). Minimum and maximum yaw models utilized the JMP screening tool, the half-normal probability plot of effects, and some trial and error. All of the tables with additional details for this modeling effort are provided in Appendix C.

Table 8. Summary of Modeling Results for Flight With Turbulence

	Max Roll Turb	Min Roll Turb	Max Pitch Turb	Min Pitch Turb	Max Yaw Turb	Min Yaw Turb	Yaw Range Turb
R^2	0.912	0.875	0.927	0.884	0.976	0.996	0.885
R^2 Adj	0.892	0.859	0.907	0.851	0.964	0.994	0.859
RMS Err	0.557	0.543	0.391	0.466	1.288	0.434	1.069
Mean	0.758	-4.784	1.020	-3.197	5.250	0.363	4.886
Model Fval	<.0001	<.0001	0.0001	<.0001	<.0001	<.0001	<.0001
LoF Fval	0.5012	0.7835	0.3333	0.8756	0.4201	0.3119	0.3847
Regression Equation Terms with p-values*	WS - <.0001	WS - <.0001	WS - <.0001	WS - <.0001	Alt - <.0001	Alt - <.0001	WS - <.0001
	Thr - .0018	Alt - .0008	Alt - .0002	Thr - .1449	WS - <.0001	WS - <.0001	RH - .0006
	Alt - .0001	Alt * Alt - .001	Thr - .0032	Alt - .0007	RH - .0939	RH - .0382	Thr - .0113
	WS * Alt - .0278		WS * WS - .0392	WS * WS - .0175	Alt * Alt - <.0001	Alt * Alt - <.0001	Alt - .0004
	Alt * Alt - .001		WS * Alt - .0834	Thr * Thr - .0518	Alt * WS - <.0001	Alt * WS - <.0001	Alt * Alt - .030
			WS * Thr - .0185	WS * Alt - .0183	RH * RH - <.0001	RH * RH - <.0001	
					Alt * Alt * Alt - .0084	Alt * Alt * Alt - <.0001	
					Alt * Alt * WS - .0015	Alt * Alt * WS - <.0001	
					Alt * Alt * Alt - .0004	Alt * Alt * Alt - <.0001	
* Alt - Altitude, RH - Relative Heading, Thr - Throttle, WS - Wind Speed							

The modeling results at this point seem reasonable with a couple of concerns. First of all, the R^2 and R^2 adjusted values seem high for a process like this with all of the values over .85. Additionally, all of the parameters are significant with very low p-values and no lack of fit. The only thing that is troubling is that with the high R^2 values and multiple high order terms for maximum and minimum yaw, there is the possibility that the model is mispecified and over fit; the concern with being over fit is noise being fit by the higher order terms. Additionally, the mean of maximum roll seems low; especially considering relative heading is not in the regression equation.

Residual Analysis

Next, the statistical assumptions were verified. The residual plots of concern are provided in Figures 7-9. Additional residual plots are given in Appendix D.

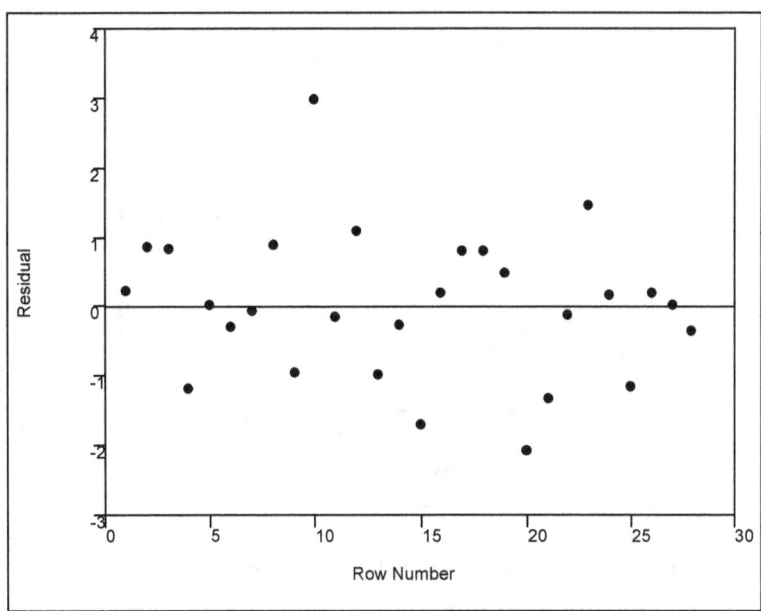

Figure 7. Plot of Max Yaw Residuals versus Row Number

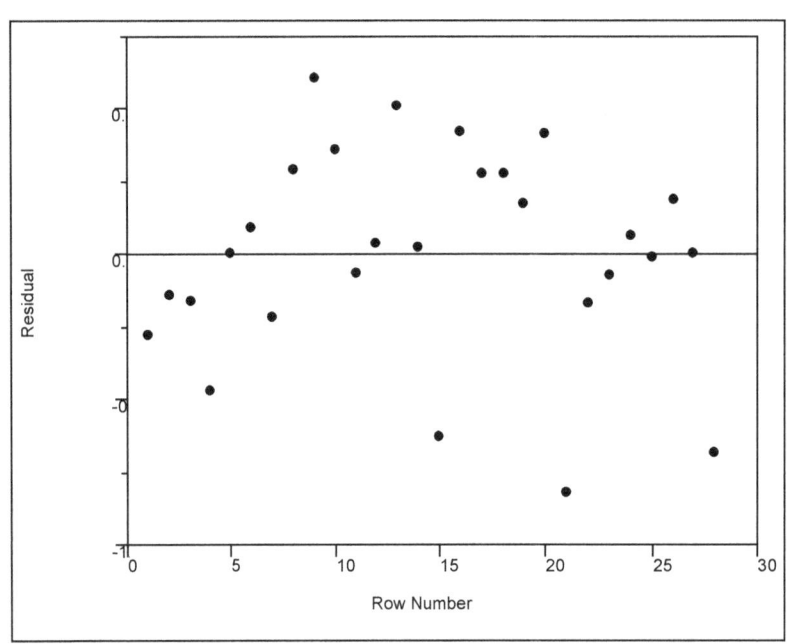

Figure 8. Plot of Min Yaw Residuals versus Row Number

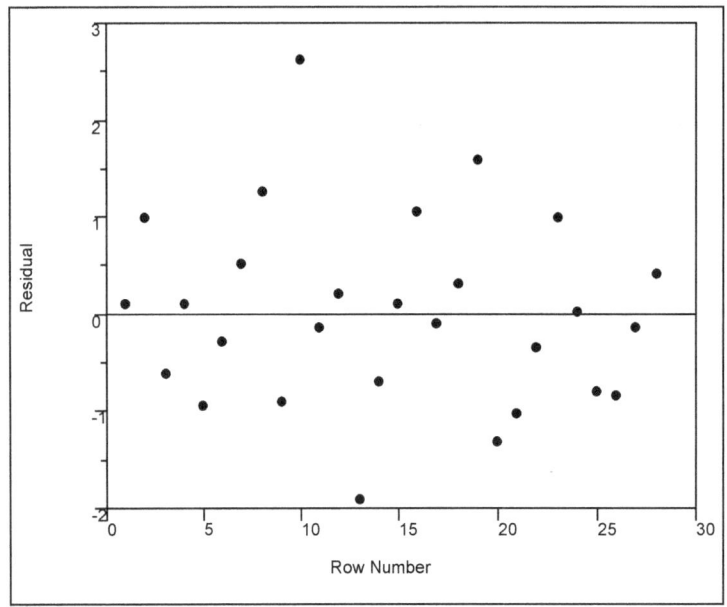

Figure 9. Plot of Yaw Range Residuals versus Row Number

The residual plots do not reveal any glaringly obvious problems with the assumption required for this methodology. None of the normal probability plots appear to show any issues with the normality assumption. Some of the residual versus predicted

36

plots showed slight hints of non-constant variance, but not enough to be too concerned. Most of the residual versus row plots look good as well; however, the maximum and minimum yaw residual versus row plots (Figures 7 & 8) show that there are clusters of residuals near one another. Again, this may be due to the fact that the aircraft's wind estimation system cannot be cleared between runs and is an estimation that is continually updated meaning that the estimation data from the previous run (or runs) is probably affecting the aircraft's yaw in each run. This problem does not seem to affect the yaw range residuals when plotted sequentially (Figure 9). Yaw range may be a better choice for a real-world implementation of this methodology.

Comparison of With and Without Turbulence Models

Tables 9 and 10 compare the models developed for the experiments with and without turbulence. Across the board, the models with turbulence are an improvement in explaining the variance of the system (R^2) as well as the fit (lack of fit values increase). However, as discussed previously, the number of factors included in the models with turbulence for minimum and maximum yaw may be unacceptably high and could be fitting noise.

Table 9. No Turbulence and Turbulence Response Comparison

	Max Roll NT	Max Roll Turb	Min Roll NT	Min Roll Turb	Max Pitch NT	Max Pitch Turb
R^2	0.776	0.912	0.781	0.875	0.801	0.927
R^2 Adj	0.737	0.892	0.743	0.859	0.701	0.907
RMS Err	0.143	0.557	0.186	0.543	0.112	0.391
Mean	-1.816	0.758	-2.513	-4.784	-0.817	1.020
Model Fval	<.0001	<.0001	<.0001	<.0001	0.0001	0.0001
LoF Fval	0.0225	0.5012	0.1607	0.7835	0.0153	0.3333
Regression Equation Terms	RH	WS	RH	WS	WS	WS
	Thr	Thr	Thr	Alt	RH	Alt
	Alt	Alt	Alt	Alt * Alt	Thr	Thr
	RH * Alt	WS * Alt	RH * Alt		Alt	WS * WS
		Alt * Alt			WS * RH	WS * Alt
					WS * Thr	WS * Thr
					RH * Thr	
					Thr * Thr	
					WS * Alt	
* Alt - Altitude, RH - Relative Heading, Thr - Throttle, WS - Wind Speed						

Table 10. With and Without Turbulence Response Comparison (cont)

	Min Pitch NT	Min Pitch Turb	Max Yaw NT	Max Yaw Turb	Min Yaw NT	Min Yaw Turb	Yaw Range
R^2	0.634	0.884	0.508	0.976	0.898	0.996	0.885
R^2 Adj	0.588	0.851	0.447	0.964	0.869	0.994	0.859
RMS Err	0.151	0.466	7.894	1.288	2.161	0.434	1.069
Mean	-1.443	-3.197	5.006	5.250	2.919	0.363	4.886
Model Fval	<.0001	<.0001	0.0006	<.0001	<.0001	<.0001	<.0001
LoF Fval	0.1565	0.8756	0.8488	0.4201	<.0001	0.3119	0.3847
Regression Equation Terms	WS	WS	WS	Alt	WS	Alt	WS
	RH	Thr	Alt	WS	RH	WS	RH
	Thr	Alt	WS * Alt	RH	Alt	RH	Thr
		WS * WS		Alt * Alt	RH * RH	Alt * Alt	Alt
		Thr * Thr		Alt * WS	WS * Alt	Alt * WS	Alt * Alt
		WS * Alt		RH * RH	Alt * Alt	RH * RH	
				Alt * Alt * Alt		Alt * Alt * Alt	
				Alt * Alt * WS		Alt * Alt * WS	
				Alt * Alt * Alt * Alt		Alt * Alt * Alt * Alt	
* Alt - Altitude, RH - Relative Heading, Thr - Throttle, WS - Wind Speed							

Tolerance Interval Methodology

The regression equations obtained from the designed experiment return point estimates for the responses; they yield expected values for the responses at certain factor settings. The tolerance interval provides the upper or lower bound for which a chosen percentage of future values would fall below or above with a chosen confidence level. This will result in a set of equations that vary based on the aircraft's current flight parameters (only the parameters significant for a given model) and an added buffer based on the chosen population percentage, confidence level and the model's standard error at the specific experimental data point. Due to the issue with the mean of maximum roll being low than expected and the regression equation not including the relative wind heading, only the maximum of the absolute value of minimum and maximum roll values at each data point will be considered in the final model. This results in an absolute value maximum roll model that is just the positive version of the minimum roll model. Additionally, due to the issues with the minimum and maximum yaw, only the yaw range will be considered in the final models. Table 11 shows the point estimates for each of the four responses at each experimental data point as well as the computed 99% population proportion, 99% confidence level upper or lower tolerance interval bounds.

Table 11. Computed Tolerance Interval Bounds for Responses

Run #	Abs Max Roll	99/99 TI	Max Pitch	99/99 TI	Min Pitch	99/99 TI	Yaw Range	99/99 TI
1	5.12	7.09	0.88	2.30	-2.84	-4.56	5.29	9.18
2	6.46	8.49	3.28	4.84	-4.86	-6.68	9.49	13.62
3	5.12	7.09	1.41	2.94	-3.29	-5.35	6.50	10.68
4	5.12	7.09	0.35	1.88	-3.87	-5.93	4.09	8.26
5	2.70	5.10	0.17	1.70	-2.08	-3.93	1.58	6.29
6	2.99	5.02	-0.31	1.25	-1.96	-3.78	2.68	6.82
7	3.85	5.87	0.01	1.57	-2.41	-4.23	1.97	6.11
8	5.61	7.63	1.19	2.75	-3.79	-5.61	6.46	10.59
9	5.61	7.63	1.19	2.75	-3.79	-5.61	5.13	9.27
10	6.46	8.49	3.28	4.84	-4.86	-6.68	8.16	12.30
11	5.12	7.09	0.88	2.30	-2.84	-4.56	6.62	10.80
12	3.85	5.87	0.04	1.61	-2.12	-3.94	3.18	7.31
13	5.12	7.09	0.88	2.30	-2.84	-4.56	5.29	9.18
14	5.12	7.09	0.88	2.30	-2.84	-4.56	5.29	9.18
15	6.46	8.49	2.25	3.81	-5.15	-6.97	8.28	12.41
16	2.99	5.02	-0.34	1.22	-2.25	-4.07	1.48	5.61
17	7.73	9.85	3.92	5.64	-5.91	-7.97	10.28	14.45
18	2.51	4.63	-0.85	0.87	-1.63	-3.69	0.31	4.49
19	2.99	5.02	-0.34	1.22	-2.25	-4.07	0.15	4.28
20	6.46	8.49	2.25	3.81	-5.15	-6.97	6.95	11.09
21	5.61	7.63	2.22	3.78	-3.50	-5.32	6.34	10.47
22	5.12	7.09	0.88	2.30	-2.84	-4.56	5.29	9.18
23	5.61	7.63	2.22	3.78	-3.50	-5.32	7.67	11.80
24	5.12	7.09	0.88	2.30	-2.84	-4.56	3.97	8.15
25	3.85	5.87	0.01	1.57	-2.41	-4.23	3.30	7.43
26	3.85	5.87	0.04	1.61	-2.12	-3.94	4.50	8.64
27	4.41	6.80	1.59	3.11	-3.60	-5.45	5.22	9.93
28	2.99	5.02	-0.31	1.25	-1.96	-3.78	1.36	5.49

Modeling the Tolerance Interval of the Responses

The computed tolerance interval data generated for each of the design points for each of the four chosen responses (maximum absolute value roll, maximum and minimum pitch and the yaw range) was then modeled using regression analysis. Since the tolerance interval is computed from a combination of the point estimate (the previously generated regression equations) and an additional component to take into consideration the variance of the data at the specific factor settings as well as the percent interval being computed, one would not expect problems with the model fit given that the same factors that were significant in the previous regression equations were utilized. Indeed, every model developed has R^2 and R^2 adjusted values of greater than 0.999 with p-values for all of the significant factors from the previous regressions at < 0.0001. Equations 8 – 11 are the 99% population proportion, 99% confidence tolerance interval equations generated.

$$\textbf{\textit{Abs Max Roll}} = 5.50 - .327 * Wind\ Speed + .0057 * Alt$$
$$- .000053 * (Alt - 180)^2 \tag{8}$$

$$\textbf{\textit{Max Pitch}} = 2.06 + .298 * Wind\ Speed - .0047 * Alt - .0177 * Throttle$$
$$- 0.00059 * (Wind\ Speed - 8)(Altitude - 180)$$
$$- 0.004156 * (Wind\ Speed - 8)(Throttle - 70) \tag{9}$$
$$+ .0145 * (Wind\ Speed - 8)^2$$

$$\textbf{\textit{Min Pitch}} = -2.657 - .268 * Wind\ Speed + .0051 * Alt - .0096$$
$$* Throttle + 0.00099$$
$$* (Wind\ Speed - 8)(Altitude - 180) \tag{10}$$
$$- 0.00112(Throttle - 70)^2 - .0187 * (Wind\ Speed - 8)^2$$

$$\textbf{\textit{Yaw Range}} = 8.00 + .623 * Wind\ Speed - .0121 * Alt + 0.0147$$
$$* Relative\ Heading - .0402 * Throttle - 5.67e \tag{11}$$
$$- 5 * (Alt - 180)^2$$

To utilize these equations (namely the yaw range equation since it is the only one with a relative heading term), relative heading must be resolved to be useful over the full 360 degrees around the aircraft. One way to do this would be to take the absolute value of the relative heading minus 360 degrees for relative headings greater than 180 degrees.

42

This utilizes the assumption that the results for 180 to 360 degrees relative heading are symmetric to the results from 0 to 180 degrees. The recommended method for utilizing the yaw range equation would be to add one-half of the yaw range to the current aircraft yaw for a maximum yaw and subtract one-half of the yaw range from the current aircraft yaw to obtain a minimum yaw value. The maximum absolute value of roll should be both added and subtracted from zero roll to account for the possible attitude variance.

With these equations and the relative heading correction, the models can be used to create an area of variability in whatever coordinate system is being used for the acoustic propagation model. The attitude variability models would add a buffer around the straight line path from the air vehicle to the expected listener position. Figure 10 is a two dimensional simplification depicting how this would work when a specific listener position is of interest. Figure 11 is a similar depiction but represents the case when an area (versus a single listener position) is of interest. In both figures, δ_1 and δ_2 are the additional parts of the acoustic source that should be propagated.

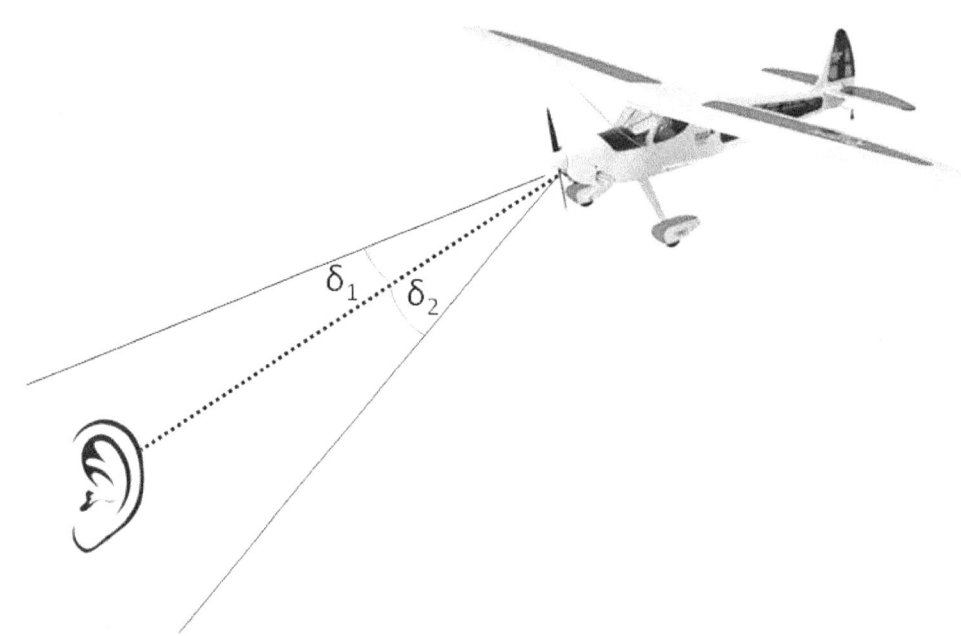

Figure 10. Depiction of Methodology With a Single Listener Position

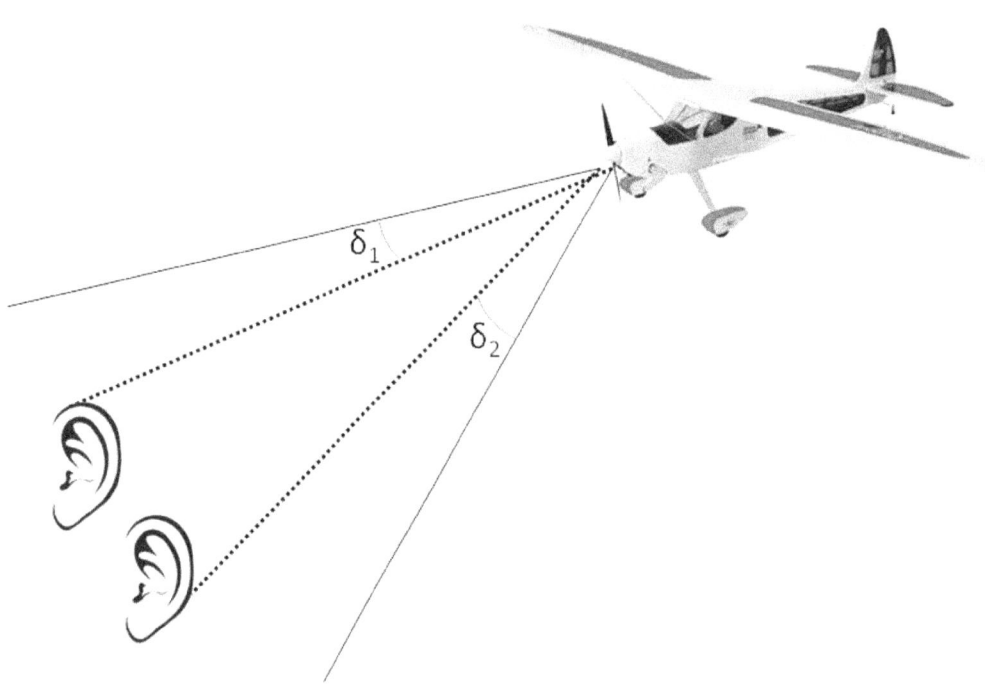

Figure 11. Depiction of Methodology With a Area of Interest

As a simplified example for how much this methodology could reduce the amount of an acoustic signature that is propagated, let us examine the case where the point of interest is directly below the aircraft. In this case, yaw is of negligible impact on the area of the acoustic signature that should be evaluated. Table 12 gives percentages of both the full sphere and the hemisphere for the simplified situation above at each of the experimental design points. Run 17 was the worst case from the experimental design points and would still provide a reduction of greater than 99.7% from propagating the full sphere and about a 98.7% reduction from propagating the lower hemisphere.

Table 12. Propagation Proportions

Run #	Deg^2 Propagated	Percent of Sphere	Percent of Hemisphere
1	97.3	0.24%	0.47%
2	195.6	0.47%	0.95%
3	117.6	0.28%	0.57%
4	110.7	0.27%	0.54%
5	57.4	0.14%	0.28%
6	50.5	0.12%	0.25%
7	68.2	0.17%	0.33%
8	127.7	0.31%	0.62%
9	127.7	0.31%	0.62%
10	195.6	0.47%	0.95%
11	97.3	0.24%	0.47%
12	65.2	0.16%	0.32%
13	97.3	0.24%	0.47%
14	97.3	0.24%	0.47%
15	183.0	0.44%	0.89%
16	53.1	0.13%	0.26%
17	268.2	0.65%	1.30%
18	42.2	0.10%	0.20%
19	53.1	0.13%	0.26%
20	183.0	0.44%	0.89%
21	139.0	0.34%	0.67%
22	97.3	0.24%	0.47%
23	139.0	0.34%	0.67%
24	97.3	0.24%	0.47%
25	68.2	0.17%	0.33%
26	65.2	0.16%	0.32%
27	116.5	0.28%	0.56%
28	50.5	0.12%	0.25%

V. Conclusions and Recommendations

Chapter Overview

This chapter concludes this research effort by discussing the results and implications of Chapter 4 and comparing the results against the research objectives as specified in Chapter 1. Further, recommendations for follow-on activities as well as possible future research are explored. Follow-on activities would be focused on validating or improving the results from this research related strictly to characterizing the attitude variance of small UAVs while future research efforts would look towards further developing acoustic dynamic mission planning tools that could be utilized for autonomous or semi-autonomous use.

Research Conclusions

The first stated research goal was to develop estimates for the minimum and maximum roll, pitch and yaw for level flight for the Sig Rascal 110. Models were developed and the models for roll and pitch seem to accurately represent the simulation process that it is modeling. However, due to the non-immediate autopilot correction for yaw (unlike roll and pitch) because it is controlled for navigation purposes and the fact that the autopilot wind estimation algorithm is continually refined based on aircraft sensor data, modeling yaw variability was more problematic. Utilizing the range methodology (rather than the minimum and maximum yaw response) offers a viable alternative to the intended methodology. However, this method is likely very conservative (as yaw does not appear to vary as rapidly as roll and pitch in turbulent environments) and may result in propagating more of the acoustic signature than

47

necessary. Maximum roll was similarly modified due to the mean of the response remaining negative despite the fact that relative heading was not included in the regression equation. Using the maximum of the absolute value of roll was used as the response instead, with this value added and subtracted from zero to define the roll variation.

These models were slightly modified to use the tolerance interval methodology specified by the second research goal. Tolerance intervals were computed for the experimental design points with 99 percent population proportion at 99 percent confidence tolerance intervals. Models were then fit using this tolerance interval data using the methodology addressed at the end of Chapter 4.

Finally, the methodology for developing models for other aircraft is discussed in Chapter 3 and will be addressed further as follow-on activities are discussed.

Follow-On Activities

As stated, one purpose of follow-on activities is to validate the models developed for characterizing the attitude variance of small UAVs, thus providing confidence in the research results. Follow-on activities could additionally be focused on improving the attitude variance models developed through this research. The following are possible areas for follow-on activities.

Real-world Validation of Attitude Variance Models

Prior to implementing the results of this research, it is imperative that it is validated by real-world experimentation. This research was conducted on the substantial assumption that results from the simulation environment (FlightGear) would translate to the real-world. If the attitude variation models are implemented in the real-world as suggested by this research under the assumption that the results translate to the real-world, it could result in a situation where the aircraft's actual attitude varies more than what the model predicts, thus possibly resulting in the aircraft being acoustically detected when it was predicted not to be able to be heard. This would be due to a portion of the aircraft's signature not being taken into consideration when it should be considered for detection. It is also not recommended that extensive additional simulation work be conducted until the results are validated. The cost could be that further man hours are spent on research that is not valid to real-world operations.

Extension to Other Platforms and Simulation Environments

This research effort could easily be utilized to characterize the attitude variance of other aircraft in a similar manner. However, (as stated above) it is highly advisable that a validation effort precede any significant effort to model other platforms. Additionally, it should be noted that "extension to other platforms" does not mean that simply using the results from this research with other platforms prior to further experimentation is advisable. That being said, there are many real-world aircraft with models available (for free) to use with the FlightGear simulation platform. However, it is unclear if any of the aircraft available with FlightGear are themselves UAVs or have been adapted to use as UAVs. Either way, most UAVs in use by the Department of Defense have an accompanying simulation environment developed to train operators. If this is the case, the simulation environment is flexible enough to change the wind speed and direction (as well as add turbulence), and the data required for this research (the responses) is available post simulation run, the experimental design utilized in this research could minimally serve as a good starting point for characterizing the attitude variance of other platforms.

Future Research

The following are some possible future research areas that could serve to further develop the results of this research

Attitude Variance Model Generalization

Model generalization could be applied within the models of a single aircraft or could be generalized across platforms. As new models are developed using this methodology, it is desirable that the results are compared against one another. It is possible that aircraft attitude variability can be predicted by factors related to the aircrafts' physical dimensions, flight characteristics, or autopilot characteristics. If models for enough platforms are developed, some of these differences may be discerned and tolerances may be made for aircraft without having to perform a full experiment with each platform (although real-world validation runs would be recommended).

Modeling Aircraft Loiter/Turn Attitude Variance

When viewed from the perspective of this research, as an aircraft turns, its heading relative to the wind direction, average roll, yaw and pitch all vary. Of those factors, the heading relative to wind direction and its pitch (minimally) will also continue to change through the turn. It is possible that the models developed here would be robust enough to deal with these changes, but regardless, the predicted responses would need to be calculated repeatedly (and quickly) to account for the rapidly changing heading relative to wind direction. It would likely be preferable to develop a model specifically for aircraft loiters of various radii (as well as various maximum aircraft bank angles). A model developed for aircraft loiters could likely be utilized momentarily for aircraft turns as well.

Other Source Reduction Techniques

With the DoD's proposals for multiple aircraft controlled by single operators, it is not likely that full, robust acoustic modeling will be available to run in real time in the near future. Therefore, it is advisable to continue to research other methodologies for reducing the portion or fidelity of the acoustic source that is modeled. One example, with respect to source fidelity, is applying the psychoacoustic phenomena of auditory masking which could result in reductions in the amount of acoustic data stored. Auditory masking occurs when a sound is made inaudible by a louder sound of similar frequency and duration [18]. Application of this concept could eliminate the need for storing and utilizing bands of quiet acoustic data that are near much louder bands of data.

Human User Interface/Autonomy Using Acoustic Data

Acoustic data is easily implemented in pre-planned mission scenarios where the aircraft is to follow one path to the target area and follow the pre-planned path to minimize acoustic detection. However, the addition of acoustic data real-time and the use of this data for dynamic mission planning is not well-studied. Research into methods in which the acoustic data is displayed to the user and how the user should utilize the data should be conducted. Additionally, the feasibility of using the acoustic data to autonomously change the flight path could also be examined.

Summary

This research utilized two central-composite designs in a simulation environment to model the minimum and maximum roll, pitch and yaw of a Sig Rascal 110 in the presence and absence of turbulence. It was determined that turbulence was necessary to perturb the aircraft and thus measure how the aircraft's autopilot responded to those perturbations. It was also determined that the minimum and maximum responses were not as useful for the aircraft's yaw as that process is governed by a control loop that does not "correct" the heading as quickly as the roll and pitch. Therefore the yaw range was used as a response instead and thus recommended for implementation. Maximum roll was similarly modified due to the mean of the response remaining negative despite the fact that relative heading was not included in the regression equation. Using the maximum of the absolute value of roll was used as the response instead, with this value added and subtracted from zero to define the roll variation.

The data set with turbulence was used to create 99 percent population proportion and 99 percent confidence tolerance intervals for the maximum of the absolute value of roll, the minimum and maximum pitch and the yaw range. These values were modeled using regression analysis and the models were used to evaluate this methodology for the case when the aircraft is directly above a single listener position. The methodology was shown to reduce the propagated acoustic signature by over 99.3% for all of experimental design points in this specific case.